# ONE HUNDRED WAYS
## TO A
### *Happy Cat*

ALSO COMPILED BY CELIA HADDON

One Hundred Ways to be Happy
One Hundred Ways to Say I Love You
One Hundred Ways to Say Thank You
One Hundred Ways to Serenity

# ONE HUNDRED WAYS
## TO A
# *Happy Cat*

### COMPILED BY
# Celia Haddon

Hodder & Stoughton
LONDON   SYDNEY   AUCKLAND

British Library Cataloguing in Publication Data
A record for this book is available from
the British Library

ISBN 0 340 74598 3

Printed and bound in Great Britain by
Clays Ltd, St Ives plc

Hodder and Stoughton
A Division of Hodder Headline Ltd
338 Euston Road
London NW1 3BH

# Contents

THE GOOD OWNER'S GUIDE     7

THE JOY OF KITTENS     20

CAT CUISINE     29

THE ART OF FELINE MAINTENANCE     40

THE INDOOR CAT     51

LITTER TRAY TROUBLES     61

YOUR CAT AND OTHERS     73

GOLDEN OLDIES     83

USEFUL INFORMATION     95

# The Good Owner's Guide

Y ou think you own your cat but the cat knows it owns you. Bear this in mind. It will make sense of the cat-human relationship.

Your cat will train you into its requirements eventually so you might as well reorganise your daily routine for its benefit now.

Do not sleep late. You will be woken well before the alarm call by your cat purring loudly, dive-bombing your intimate bits, biting your toes under the duvet, or swiping items off the dressing-table. Cats do not recognise weekends.

**B**an small girls from the house. Small girls are tempted to commit that ultimate crime – dressing up cats.

Give your cat an occasional high. Grow and dry your own catnip (*Nepeta cataria*). Plant the kolomite vine (*Actnidia kolomikta*), valerian (*Valeriana officinalis*) and spider plant (*Chlorophytum comosum* 'Vittatum').

A yawn is a cat's honest opinion. Some cats just say no to sniffing or eating drugs.

A ncestral feline tradition from the days before cats domesticated humans means that a cat likes to move from one bed to another, preferably yours. Provide more than one bed per cat.

Be a conscientious door attendant. Cats want to be in when they are out, out when they are in, and sometimes both simultanously.

Leave all cupboards and drawers open. If you persist in closing them, you will have check each time to see if a cat is inside.

To reduce the frequency of feathers in the living-room and tiny corpses on the pillow, keep your cat indoors at night. This will also save your cat from being run over in the dark.

Supply a scratching post and express your astonishment and pleasure with a food reward if by chance your cat condescends to use it.

D raw vertical lines on the scratching post with a felt-tipped pen. Vertical lines sometimes prompt competitive scratching.

If your cat persists in expressing its inner kitten by scratching on the furniture, apply double-sided sticky tape.

To confirm that your cat is two-timing you with a neighbour, put a paper collar round its neck, fixed with a paperclip or sticky tape, with a clear message asking the finder to call you.

# The Joy of Kittens

Start training your kitten early. It has already started training you.

When you take your kitten home for the first time, use the same kind of litter in the tray as it has been used to in the breeder's home.

Give your kitten a confident purrsonality by getting it used to dogs, other adult cats, the noise of household machines, TV noises, and children in its early weeks. An overprotected kitten makes a nervous cat.

What a kitten eats, the cat will eat. Feed your kitten a variety of food so that it doesn't grow up to be a finicky cat. Well, that's the theory anyway.

If you want your kitten to become accustomed to car travel, drive it around from as early an age as possible as often as possible.

Claw enforcement starts in kittenhood. Get your kitten used to having its nails clipped regularly.

Get your kitten used to being brushed. It will still shed fur on your smartest outfit, but a little less fur means a more refined pattern.

Tomcats get into fights, roam away from home, leave offensive smells, and join noisy gangs. Neuter your male kitten. He will never know what he missed.

Spay your female kitten. Would you want to be responsible for 20,000 descendants in five years?

# Cat Cuisine

Vary the menu. Cats like variety. Stolen food, from your or the other cat's plate, tastes sweetest.

Turn your kitchen into a snack bar. Serve several small meals a day, rather than just one or two large ones. Cats like to snack, not binge.

Is your cat obese? Stop being an enabler. Give the same diet food every day. Measure it out. No little extras. Don't cheat. But you know all this, don't you?

The purrfect feline meal has the temperature of a fresh dead mouse, so don't take cat food directly out of the fridge.

Warm up tinned food to tempt a picky eater.

To prevent vomit landing in your lap on Christmas Day, keep the turkey and the pudding secure.

Ordinary tap water? You must be joking. Cats don't like the taste of chemicals. Give your cat clean rainwater or still mineral water.

Let your taps drip if your cat enjoys a drink from a running tap – the motion of the water seems to reduce the chemical smell of tap water.

Encourage your cat to drink more, if it's fed dried food, by serving special milk sold solely for cats, or water flavoured with fish juices or meat juices.

Keep fresh water always available. Cow's milk is bad for some cats who cannot digest the lactose. Buy the reduced-lactose milk sold specially for cats.

Tinned sardines or pilchards in tomato sauce (yes, tomato sauce) fed by hand, mouthful by mouthful, will tempt a sick cat to eat.

For once, don't be manipulated by your cat. Cats adore liver but it can cause vitamin A poisoning. Feed it only occasionally.

Your cat is not a dog, whatever it thinks. Do not let it eat dog food. This does not have the ingredients a cat needs.

Some cats, like humans, roam. Feed your feline four times a day at set times, in the hope it will come home for food, if not for love.

# The Art of
# Feline
# Maintenance

Enrich a vet. Give your cat an annual health check-up – teeth, ears, fur and fleas.

Cats with white faces or ears are more vulnerable to skin cancer. Keep them indoors for a noon nap. Or put children's sun barrier cream on the white skin.

To give a cat a pill, kneel and place your cat facing away from you between your knees. With thumb and middle finger, pull cat's head back until it is facing straight up at the ceiling at 90 degrees. Open the cat's mouth with gentle pressure from thumb and middle finger on either side of the cat's jaw. Pop tablet on the tongue as far back down the throat as possible. Close cat's mouth and keep the head pointing up at the ceiling. And the best of luck.

Coating a pill with a yeast
extract spread helps dis-
guise the taste.

Remember your cat in your will, just in case of your sudden death. If you tell your cat you've left it a legacy, it might start being nicer to you.

Get flea treatment for your cat. Remember the fleas will bite you too.

If your cat starts pulling its fur out, it may be allergic to fleas. Treat the house with an anti-flea spray and treat the cat with a flea remedy from your vet. If this doesn't help, go for behaviour counselling.

If your cat refuses to eat, see a vet. Self-starvation can lead to serious illness.

Brush your cat regularly to prevent hairballs. You are less likely to put your foot on vomit when you get out of bed first thing in the morning.

Protect cats from poisons like aspirin, do-it-yourself materials, and household disinfectants, cleaners and polishes, and fumes from over-heated pans.

Provide an indoor lawn for your cat to nibble instead of poisonous house plants like busy Lizzie (*Impatiens wallerana*), poinsettia (*Euphorbia pulcherrima*), morning glory (*Ipomoea tricolor*), dumb cane (*Dieffenbachia*), and ivy (*Hedera helix*).

Lock up wood-treatment liquids, slug pellets, weed-killers, car anti-freeze and other outdoor chemicals – your cat's curiosity might kill it.

**B**ad breath may mean bad teeth. Ask the vet to check them.

# The Indoor Cat

Keep two cats instead of one, preferably siblings from the same litter or two older cats who are already friends.

It must be a single cat? Adopt an elderly cat from a rescue shelter. Elderly cats are less active and rescue shelters find it difficult to rehome them.

Cats do pounces, not walkies. A cat in the wild makes thirty pounces a day to catch prey. Give your indoor cat thirty daily pounces using a toy fishing-rod or a piece of string. Then watch it sneer at your efforts.

Encourage ornithology as a hobby, with a bird-table within sight. Try a feeder which is attached to your window-glass, to give your cat a closer thrill.

Enrol your cat into neighbourhood watch. Provide sitting places at the window with the best view of passers-by, dogs, wildlife, other cats.

**B**uy a special wildlife video for cats, and play it while you are out.

**H**ide dry cat food so your cat has to hunt for it round the house. It makes a change from plundering the trash can or butter dish.

Punch holes in a yoghurt tub with a firm lid and put in dry food. Your cat should push the tub around until the food falls out. It may decide being fed this way is demeaning.

Change toys daily, so that your cat doesn't get bored. You can store them where the cat has left them, underneath the fridge.

58

Confirm your cat's natural feeling of superiority by making walkways above ground level and on high ledges where your feline friend can sit and look down on you.

Enrich your cat's life by teaching it tricks. Wait till your cat does what you want, then reward it. This is called instrumental learning, and your cat is already using this technique on you.

# Litter Tray Troubles

Do not suddenly change the type of litter. Cats like the feel of familiar litter beneath their feet. Mix the new litter with the old little by little until the changeover is complete.

Cats dislike change. If you must reposition the litter tray, move it only a few inches a day to its new place.

Put the litter tray in a secluded place and supply a covered tray, if your cat prefers it. Cats feel insecure while on the litter tray.

Reduce feline family stress by making sure there are enough litter trays – one per cat.

Very refined cats sometimes require two litter trays – one for one thing, the other for the other.

Clean the litter tray at least twice a day. You wouldn't like a dirty bathroom, would you?

Buy expensive small-grained litter rather than the cheaper large-grained brands. Cats have simple requirements – simply the best.

Cats who go outside still need a litter tray – for wet days, for days when they are ill, or for protection from the neighbouring feline bully. Would you want to have to go in the cold and wet?

If long-haired cats shed tell-tail litter grains or worse when they step out of the litter tray, trim the fur under their tails.

Cats who leave their bottom outside of the entrance into the covered litter need a smaller tray inside a larger one with a gap of about two and a half inches at the entrance. This ensures that the cat has to step across the gap to enter and use the inner try. Litter in the gap absorbs spills.

Perfumed household cleaners or disinfectants to clean up cat mess often smell like urine to cats and they mark over them with more urine. So don't use these to clean up cat mess.

Clean up cat mess with biological washing liquid or surgical spirit, and dry with a hairdryer.

Give your cat help, not blame, if it messes in the wrong place. It is making its mark on the world to make itself feel less stressed. Call in a cat behaviour counsellor.

# *Your Cat and Others*

I f your new partner and your cat won't share you, rehome it – the partner, of course.

Help your cats stay friends by installing more than enough cat beds, food bowls and litter trays, thus diminishing competition.

Dinner together promotes friendship. Slowly move the food bowls closer to encourage feline harmony.

Increase home security with an electronic catflap which opens only to your own cat.

Help your cat patrol its territory by ambushing unwanted feline intruders with a powerful long-distance water pistol. All's furr in love and war.

Discourage cat burglars in search of your cat's dinner from entering the house by putting oil of olbas on the outside of the catflap.

One cat leads to another. A resident cat will accept a young kitten better than a new adult, and one of the opposite better than the same sex.

Introduce a new cat in an indoor pen with litter tray, food and water for a few days, where existing cats can meet it. Swap bedding between cat beds and transfer soiled litter from one litter tray to another, to mix their smells.

If one cat is being bullied by another, install an electronic catflap into a spare cupboard or room so that the victim has its own retreat.

If two cats in the same household seriously fall out, try reintroducing them as if one of them was entirely new.

Hostile felines, passing dogs or wild animals seen through a window or a glass door can upset a nervous cat. Block off the sight of these to help your cat's peace of mind. Otherwise it may resort to the ancient art of spraying.

Make an outdoor latrine in a dry place in your own garden, using soiled litter mixed with peat or clean builders' sand. Hostile neighbouring gardeners may harm your intruding cat.

# Golden Oldies

Don't be selfish – share your bed. Make sure all cat beds are out of draughts and in warm places.

Buy your elderly cat an electric blanket or heat pad. You may keep more of your bed for yourself.

Buy a bean bag, a thick feline duvet or a radiator hammock to give comfort to arthritic feline limbs.

Don't expect your cat to keep awake. Elderly cats sleep 75 per cent of the time. If your cat wants exercise, it will wake you – at 3 a.m.

Be your cat's hairdresser and brush it daily. Elderly cats cannot always groom themselves properly.

Be your cat's manicurist. Older cats may not wear down their claws as much as they used to.

Build or buy a cat ramp to help your cat get on to your favourite armchair.

A dapt a low kitchen tray to take litter. An arthritic cat may find it difficult to climb into a high-sided litter tray.

Be your cat's bodyguard outside the house. Old cats may not be able to push open the catflap and may have difficulty coping with hostile neighbouring moggies.

Play a daily gentle game with your golden oldie – hide and purr, hunt the feather or pawball. It is a way of checking on its mental and physical health.

A gentle five-minute body massage is purrfect therapy for your older cat.

Blind or partially sighted cats need their food, litter and furniture kept in the same place. So don't leave unexpected obstacles in their path.

Deaf cats can learn hand signals for instructions or the flashing of a torch to call them in for meals. Some deaf cats can 'hear' the vibrations caused by a sharp clap of the hands. Most also 'hear' the vibration of a tin being opened.

Give your cat the last gift of all, the blessing of an easy death.

A proportion of the royalties of this book will go to:

The Feline Advisory Bureau
Taeselbury
High Street
Tisbury
Wiltshire
SP3 6LD
Helpline: 01747 871872
Fax: 871873

# Useful Information

Association of Pet Behaviour Counsellors. PO Box 46, Worcester WR8 9YS. 01386 751151. E-mail: Apbc@petbcent.demon.co.uk
Website: www.apbc.org.uk

Heated beds from Cozee Comfort heated pads, Burco Dean Appliances, Rosegrove, Burnley, Lancs, BB12 6AL. 01282 422851. Or Petnap heaters, Forder Green Farm, Woodland, Ashburton, Devon TQ13 7LP. 01364 652533

Ramps for ageing dogs and cats from Petlife, 01284 754256. Petlife also make Vetbed, simulated fur bedding, and Flectabed thermal bedding.
Website: www.vetbed-uk.demon.co.uk

Zoomgroom cat rubber brushes, which take out dead hair, are obtainable from The Company of Animals, PO Box 23, Chertsey, Surrey KT16 0PU. 01932 566696

Double-sided sticky tape, Sticky Paws. You can find this in household catalogues such as Scotts, 01285 653153. Or contact Sticky Paws, 1320 Lake St, Forth Worth, TX76102, USA, for UK stockists.
Website: www.cats4sale.com/cats/sticky_paws.htm

Staywell electronic pet doors are in most pet shops. If not ring 01772 793793.

Play'n'Treat cat balls to amuse indoor cats should be in pet shops. In case of difficulty contact Underworld Products, Belton Rd West, Loughborough, LE11 5TR. 01509 610310.
Website: www.underworldproducts.co.uk

Cat aerobic centres are found in most pet shops but big ones may need to be ordered direct from Sinclair Pet Accessories Ltd, Becks Mill, Westbury Leigh, Westbury, Wilts, BA13 3SD. 01373 864775

Pet mail order catalogue: Pets Pyjamas, 0800 783 0440.
Webite: www.petspyjamas.com